Math Workbooks 3rd Grade: Fractions

(Baby Professor Learning Books)

SPEEDY PUBLISHING

Speedy Publishing LLC
40 E. Main St. #1156
Newark, DE 19711
www.speedypublishing.com

Copyright 2015

All Rights reserved. No part of this book may be reproduced or used in any way or form or by any means whether electronic or mechanical, this means that you cannot record or photocopy any material ideas or tips that are provided in this book

Comparing Proper Fractions

Compare the fractions, and write > or < or = between them.

1. $\frac{2}{3}\ \square\ \frac{4}{8}$

2. $\frac{3}{6}\ \square\ \frac{1}{4}$

3. $\frac{3}{6}\ \square\ \frac{4}{8}$

4. $\frac{1}{4}\ \square\ \frac{2}{6}$

5. $\frac{1}{3}\ \square\ \frac{2}{5}$

6. $\frac{4}{7}\ \square\ \frac{1}{5}$

7. $\dfrac{4}{8}\ \square\ \dfrac{2}{3}$

8. $\dfrac{3}{7}\ \square\ \dfrac{1}{5}$

9. $\dfrac{2}{3}\ \square\ \dfrac{3}{8}$

10. $\dfrac{2}{5}\ \square\ \dfrac{4}{8}$

11. $\dfrac{2}{8}\ \square\ \dfrac{4}{7}$

12. $\dfrac{3}{7}\ \square\ \dfrac{4}{5}$

13. $\dfrac{2}{3}\ \square\ \dfrac{4}{5}$

14. $\dfrac{7}{8}\ \square\ \dfrac{3}{4}$

15. $\dfrac{4}{7}\ \square\ \dfrac{6}{8}$

16. $\dfrac{1}{3}\ \square\ \dfrac{3}{7}$

17. $\dfrac{5}{7}\ \square\ \dfrac{2}{6}$

18. $\dfrac{2}{8}\ \square\ \dfrac{3}{5}$

19. $\dfrac{2}{5}\ \square\ \dfrac{6}{7}$

20. $\dfrac{2}{4}\ \square\ \dfrac{1}{3}$

21. $\dfrac{3}{7}\ \square\ \dfrac{5}{8}$

22. $\dfrac{4}{8}\ \square\ \dfrac{3}{4}$

23. $\dfrac{2}{6}\ \square\ \dfrac{5}{7}$

24. $\dfrac{5}{6}\ \square\ \dfrac{6}{7}$

25. $\dfrac{3}{8}\ \square\ \dfrac{6}{7}$

26. $\dfrac{4}{6}\ \square\ \dfrac{3}{5}$

27. $\frac{4}{7}\ \square\ \frac{1}{4}$

28. $\frac{5}{8}\ \square\ \frac{3}{7}$

29. $\frac{2}{5}\ \square\ \frac{1}{3}$

30. $\frac{3}{7}\ \square\ \frac{1}{4}$

31. $\frac{4}{6}\ \square\ \frac{1}{3}$

32. $\frac{5}{7}\ \square\ \frac{4}{8}$

33. $\frac{7}{8}\ \square\ \frac{4}{7}$

34. $\frac{4}{8}\ \square\ \frac{1}{6}$

35. $\frac{1}{5}\ \square\ \frac{4}{8}$

36. $\frac{5}{8}\ \square\ \frac{3}{6}$

37. $\dfrac{3}{6}\ \square\ \dfrac{6}{7}$

38. $\dfrac{3}{4}\ \square\ \dfrac{1}{3}$

39. $\dfrac{5}{8}\ \square\ \dfrac{3}{5}$

40. $\dfrac{4}{6}\ \square\ \dfrac{3}{4}$

41. $\dfrac{2}{6}\ \square\ \dfrac{4}{5}$

42. $\dfrac{5}{7}\ \square\ \dfrac{4}{5}$

43. $\dfrac{5}{6}\ \square\ \dfrac{2}{4}$

44. $\dfrac{4}{7}\ \square\ \dfrac{7}{8}$

45. $\dfrac{3}{6}\ \square\ \dfrac{1}{4}$

Comparing Improper Fractions

Compare the fractions, and write > or < or = between them.

1. $\dfrac{5}{4}$ ☐ $\dfrac{6}{7}$

2. $\dfrac{4}{7}$ ☐ $\dfrac{2}{6}$

3. $\dfrac{1}{5}$ ☐ $\dfrac{4}{6}$

4. $\dfrac{5}{6}$ ☐ $\dfrac{1}{3}$

5. $\dfrac{5}{4}$ ☐ $\dfrac{7}{5}$

6. $\dfrac{3}{7}$ ☐ $\dfrac{4}{6}$

7. $\dfrac{3}{5}\ \square\ \dfrac{4}{8}$

8. $\dfrac{6}{7}\ \square\ \dfrac{3}{6}$

9. $\dfrac{5}{7}\ \square\ \dfrac{6}{5}$

10. $\dfrac{5}{7}\ \square\ \dfrac{3}{5}$

11. $\dfrac{7}{3}\ \square\ \dfrac{8}{4}$

12. $\dfrac{7}{6}\ \square\ \dfrac{4}{5}$

13. $\dfrac{5}{4}\ \square\ \dfrac{7}{8}$

14. $\dfrac{6}{5}\ \square\ \dfrac{5}{7}$

15. $\dfrac{5}{3}\ \square\ \dfrac{8}{5}$

16. $\dfrac{3}{2}\ \square\ \dfrac{4}{3}$

17. $\dfrac{4}{7}\ \square\ \dfrac{3}{5}$

18. $\dfrac{4}{3}\ \square\ \dfrac{7}{6}$

19. $\dfrac{7}{4}\ \square\ \dfrac{2}{1}$

20. $\dfrac{2}{3}\ \square\ \dfrac{6}{7}$

21. $\dfrac{2}{5}\ \square\ \dfrac{1}{6}$

22. $\dfrac{2}{6}\ \square\ \dfrac{1}{8}$

23. $\dfrac{3}{5}\ \square\ \dfrac{1}{8}$

24. $\dfrac{5}{6}\ \square\ \dfrac{8}{7}$

25. $\dfrac{2}{3}\ \square\ \dfrac{5}{6}$

26. $\dfrac{8}{4}\ \square\ \dfrac{5}{2}$

27. $\dfrac{2}{8}\ \square\ \dfrac{4}{7}$

28. $\dfrac{7}{8}\ \square\ \dfrac{2}{5}$

29. $\dfrac{6}{8}\ \square\ \dfrac{2}{3}$

30. $\dfrac{1}{4}\ \square\ \dfrac{2}{8}$

31. $\dfrac{3}{5}\ \square\ \dfrac{7}{8}$

32. $\dfrac{8}{7}\ \square\ \dfrac{7}{8}$

33. $\dfrac{4}{7}\ \square\ \dfrac{1}{3}$

34. $\dfrac{2}{4}\ \square\ \dfrac{4}{6}$

35. $\dfrac{7}{5}\ \square\ \dfrac{8}{6}$

36. $\dfrac{5}{8}\ \square\ \dfrac{2}{3}$

37. $\dfrac{6}{4}\ \square\ \dfrac{8}{7}$

38. $\dfrac{1}{3}\ \square\ \dfrac{3}{6}$

39. $\dfrac{2}{1}\ \square\ \dfrac{7}{3}$

40. $\dfrac{4}{8}\ \square\ \dfrac{2}{3}$

41. $\dfrac{1}{3}\ \square\ \dfrac{2}{8}$

42. $\dfrac{3}{7}\ \square\ \dfrac{5}{6}$

43. $\dfrac{5}{6}\ \square\ \dfrac{3}{8}$

44. $\dfrac{1}{5}\ \square\ \dfrac{2}{7}$

45. $\dfrac{1}{8}\ \square\ \dfrac{3}{6}$

Adding Proper Fractions

Solve.

1. $\dfrac{7}{3} + \dfrac{5}{3} =$

2. $\dfrac{8}{6} + \dfrac{3}{6} =$

3. $\dfrac{2}{5} + \dfrac{4}{5} =$

4. $\dfrac{11}{12} + \dfrac{10}{12} =$

5. $\dfrac{11}{10} + \dfrac{6}{10} =$

6. $\dfrac{7}{3} + \dfrac{1}{3} =$

7. $\dfrac{10}{3} + \dfrac{4}{3} =$

8. $\dfrac{7}{6} + \dfrac{1}{6} =$

9. $\dfrac{1}{2} + \dfrac{11}{2} =$

10. $\dfrac{10}{3} + \dfrac{10}{3} =$

11. $\dfrac{8}{6} + \dfrac{1}{6} =$

12. $\dfrac{10}{9} + \dfrac{2}{9} =$

13. $\dfrac{1}{5} + \dfrac{1}{5} =$

14. $\dfrac{2}{6} + \dfrac{7}{6} =$

15. $\dfrac{5}{7} + \dfrac{1}{7} =$

16. $\dfrac{4}{10} + \dfrac{6}{10} =$

17. $\dfrac{3}{10} + \dfrac{9}{10} =$

18. $\dfrac{10}{3} + \dfrac{11}{3} =$

19. $\dfrac{5}{10} + \dfrac{4}{10} =$

20. $\dfrac{7}{10} + \dfrac{8}{10} =$

Adding Fractions with Pie Charts

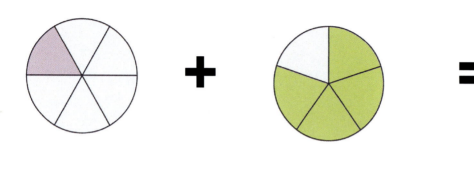

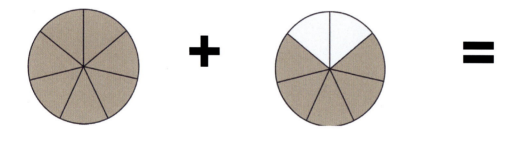

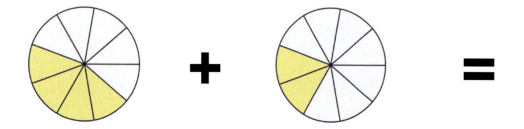

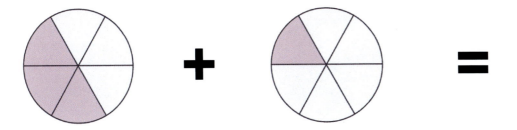

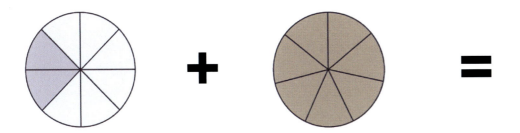

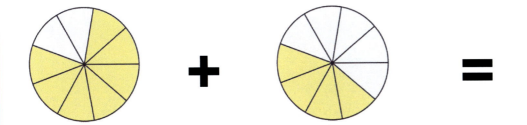

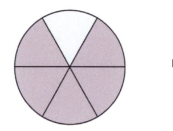 + 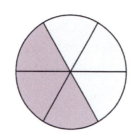 =

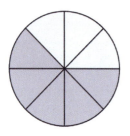 + 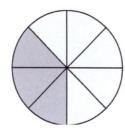 =

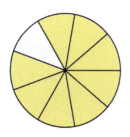  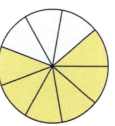 =

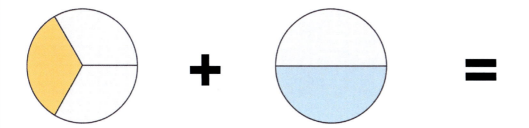

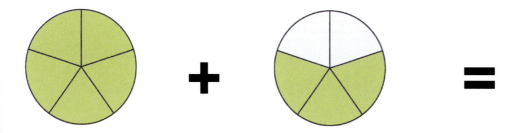

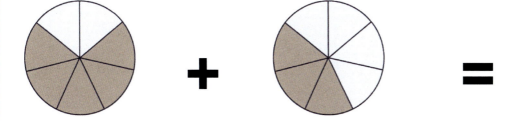

Subtracting Proper Fractions

Solve.

1. $\dfrac{4}{8} - \dfrac{1}{8} =$

2. $\dfrac{5}{8} - \dfrac{2}{8} =$

3. $\dfrac{6}{10} - \dfrac{5}{10} =$

4. $\dfrac{11}{4} - \dfrac{11}{4} =$

5. $\dfrac{8}{7} - \dfrac{4}{7} =$

6. $\dfrac{7}{2} - \dfrac{5}{2} =$

7. $\dfrac{9}{4} - \dfrac{1}{4} =$

8. $\dfrac{9}{6} - \dfrac{5}{6} =$

9. $\dfrac{3}{6} - \dfrac{2}{6} =$

10. $\dfrac{12}{3} - \dfrac{11}{3} =$

11. $\dfrac{11}{9} - \dfrac{7}{9} =$

12. $\dfrac{11}{6} - \dfrac{4}{6} =$

13. $\dfrac{8}{12} - \dfrac{8}{12} =$

14. $\dfrac{11}{10} - \dfrac{3}{10} =$

15. $\dfrac{12}{8} - \dfrac{3}{8} =$

16. $\dfrac{7}{11} - \dfrac{4}{11} =$

17. $\dfrac{1}{9} - \dfrac{1}{9} =$

18. $\dfrac{10}{7} - \dfrac{8}{7} =$

19. $\dfrac{8}{7} - \dfrac{4}{7} =$

20. $\dfrac{11}{4} - \dfrac{7}{4} =$

21. $\dfrac{12}{11} - \dfrac{12}{11} =$

22. $\dfrac{7}{2} - \dfrac{5}{2} =$

23. $\dfrac{9}{12} - \dfrac{7}{12} =$

24. $\dfrac{9}{3} - \dfrac{1}{3} =$

25. $\dfrac{12}{6} - \dfrac{3}{6} =$

26. $\dfrac{10}{3} - \dfrac{4}{3} =$

27. $\dfrac{4}{5} - \dfrac{1}{5} =$

28. $\dfrac{5}{11} - \dfrac{5}{11} =$

29. $\dfrac{12}{10} - \dfrac{2}{10} =$

30. $\dfrac{12}{10} - \dfrac{11}{10} =$

Fractions to Decimals Conversion

Solve.

1. 33/50

2. 3 4/5

3. 7 47/50

4. 9/125

5. 1 49/50

6. 13/20

7. 1/10

8. 91/100

9. 107/1000

10. 14 9/10

11. 41 1/2

12. 197/500

13. 401/500

14. 1/2

15. 7/8

16. 167/1000

17. 333/1000

18. 833/1000

19. 1/8

20. 5/8

21. 667/1000

22. 4/5

23. 2/5

24. 3/5

25. 1/4

26. 667/1000

27. 3/4

28. 1/5

29. 333/1000

30. 5/16

Score Sheet

Comparing Proper Fractions /45

Comparing Improper Fractions /45

Adding Proper Fractions /20

Adding Fractions with Pie Charts /15

Subtracting Proper Fractions /30

Fractions to Decimal Conversion /30

TOTAL /185

Congratulations!

Completion Certificate

is awarded to

_____

1. >
2. >
3. =
4. <
5. <
6. >
7. <
8. >
9. >
10. <
11. <
12. <
13. <
14. >
15. <
16. <
17. >
18. <
19. <
20. >
21. <
22. <
23. <

24. <
25. <
26. >
27. >
28. >
29. >
30. >
31. >
32. >
33. >
34. >
35. <
36. >
37. <
38. >
39. >
40. <
41. <
42. <
43. >
44. <
45. >

1. >
2. >

3. <
4. >

5. <
6. <
7. >
8. >
9. <
10. >
11. >
12. >
13. >
14. >
15. >
16. >
17. <
18. >
19. <
20. <
21. >
22. >
23. >
24. <
25. <

26. <
27. <
28. >
29. >
30. =
31. <
32. >
33. >
34. <
35. >
36. <
37. >
38. <
39. <
40. <
41. >
42. <
43. >
44. <
45. <

1. 4
2. 11/6 or 1 5/6
3. 6/5 or 1 1/5
4. 21/12 or 1 3/4
5. 17/10 or 1 7/10
6. 8/3 or 2 2/3

7. 14/3 or 4 2/3
8. 8/6 or 1 1/3
9. 12/2 or 6
10. 20/3 or 6 2/3
11. 9/6 or 1 1/2
12. 12/9 or 1 1/3
13. 2/5

14. 9/6 or 1 1/2
15. 6/7
16. 1
17. 12/10 or 1 1/5
18. 7
19. 9/10
20. 15/10 or 1 1/2

1. 1/4 + 2/3 = 11/12
2. 4/6 + 2/6 = 6/6 or 1
3. 4/8 + 2/8 = 6/8 or 3/4
4. 1/6 + 4/5 = 29/30
5. 7/7 + 5/7 = 12/7 or 1 5/7
6. 4/9 + 2/9 = 6/9 or 2/3
7. 3/6 + 1/6 = 4/6 or 2/3
8. 2/8 + 7/7 = 5/4 or 1 1/4

9. 7/9 + 4/9 = 11/9 or 1 2/9
10. 5/6 + 3/6 = 8/6 or 1 1/3
11. 5/8 + 3/8 = 8/8 or 1
12. 8/9 + 6/9 = 14/9 or 1 2/3
13. 1/3 + 1/2 = 5/6
14. 5/5 + 3/5 = 8/5 or 1 3/5
15. 5/7 + 3/7 = 8/7 or 1 1/7

1. 1. 3/8
2. 2. 3/8
3. 3. 1/10
4. 4. 0
5. 5. 3/7
6. 6. 1
7. 7. 2

8. 8 2/3
9. 9. 1/6
10. 10. 1/3
11. 11. 4/9
12. 12. 7/6 or 1 1/6
13. 13. 0
14. 14. 4/5

15.	15. 9/8 or 1 1/8	23.	23. 1/6
16.	16. 3/11	24.	24. 8/3 or 2 2/3
17.	17. 0	25.	25. 9/6 or 1 1/2
18.	18. 2/7	26.	26. 2
19.	19. 4/7	27.	27. 3/5
20.	20. 1	28.	28. 0
21.	21. 0	29.	29. 1
22.	22. 1	30.	30. 1/10

1.	0.66	17.	0.333
2.	3.8	18.	0.833
3.	7.94	19.	0.125
4.	0.072	20.	0.625
5.	1.98	21.	0.667
6.	0.65	22.	0.8
7.	.1	23.	0.4
8.	0.91	24.	0.6
9.	0.107	25.	0.25
10.	14.9	26.	0.667
11.	41.5	27.	0.75
12.	0.394	28.	0.2
13.	0.802	29.	0.333
14.	0.5	30.	0.3125
15.	0.875		
16.	0.167		

Made in the USA
Las Vegas, NV
13 March 2022